HOW STEM BUILT THE ROMAN EMPIRE

XINA M. UHL

Rosen YA
New York

For Brandy and Dave, traveling companions in Rome and life itself

Published in 2020 by The Rosen Publishing Group, Inc.
29 East 21st Street, New York, NY 10010

Copyright © 2020 by The Rosen Publishing Group, Inc.

First Edition

All rights reserved. No part of this book may be reproduced in any form without permission in writing from the publisher, except by a reviewer.

Library of Congress Cataloging-in-Publication Data

Names: Uhl, Xina M., author.
Title: How STEM built the Roman empire / Xina M. Uhl.
Description: First edition. | New York : Rosen Publishing, 2020. | Series: How STEM built empires | Includes bibliographical references and index.
Identifiers: LCCN 2019015812| ISBN 9781725341531 (library bound) | ISBN 9781725341524 (pbk.)
Subjects: LCSH: Building—Rome—History—Juvenile literature. | Technology—Rome—History—Juvenile literature. | Science—Rome—History—Juvenile literature. | Science, Ancient—Juvenile literature. | Rome—History—Juvenile literature.
Classification: LCC T16 .U46 2020 | DDC 609.37—dc23
LC record available at https://lccn.loc.gov/2019015812

Manufactured in China

On the cover: Among all of Rome's incredible architecture, Roman aqueducts are perhaps the most iconic.

CONTENTS

INTRODUCTION	4
CHAPTER ONE THE SOLDIERS	7
CHAPTER TWO THE MEASURERS	16
CHAPTER THREE THE TRAVELERS	25
CHAPTER FOUR THE BUILDERS	32
CHAPTER FIVE THE FARMERS	40
CHAPTER SIX THE HELPERS	47
CHAPTER SEVEN THE IMMORTALS	57
TIMELINE	64
GLOSSARY	66
FOR MORE INFORMATION	68
FOR FURTHER READING	71
BIBLIOGRAPHY	72
INDEX	76

INTRODUCTION

Rome is sometimes called the Eternal City for good reason. The ancient Romans established a republic in 509 BCE that lasted until the last emperor fell to invaders in 476 CE. Over its thousand-year history, Rome came to dominate the Mediterranean region, the Middle East, and Europe as far north as Britain, an area nearly 2 million square miles (5.18 million square kilometers) at its greatest extent in 117 CE. When the city of Rome was established, it had a population of only a few thousand. By the sixth century BCE, that number had grown to around twenty-five thousand. After another few centuries—around 100 BCE—between 1.5 million and two million slaves alone lived under Rome's dominion. By 14 CE, the empire's population had increased to nearly five million.

The ancient Romans took pride in their farming roots. As Roman statesman Cicero claimed in *Cicero's Three Books of Offices*: "For of all gainful professions, nothing is better, nothing more pleasing, nothing more delightful, nothing better becomes a well-bred man than agriculture." Even at the height of Roman civilization, 80 percent of the people lived outside of cities in rural estates or small farms.

Romans believed that the occupation of farming made them honorable and strong. They took great pride in their citizenship, thinking of the family, the group, and the state above their individual needs and desires. More than anything, though, the average Roman was practical.

As admirers (and conquerors) of the Greeks who came before them, the Romans adopted many components of Greek culture, including their advances in science, technology, engineering, and mathematics (STEM). However, where the Greeks preferred to dwell on philosophy and theories of mathematics and science, the Romans had little interest in such practices. To the practical Romans, STEM had no value unless it could be applied to the physical world.

The Roman Empire reached its greatest extent in 117 CE under Emperor Trajan. Such immense conquest was possible only because of the best and brightest Roman STEM innovators.

For instance, what use was physics unless it could result in better catapults or other weapons of war? Why study biology unless it could improve crop yields or produce stronger, healthier breeds of livestock? Geometry and mathematics had little worth unless they could be used to build spectacular arches, domes, and other structures.

To those ends, the Romans adopted Greek inventions and ideas and made them better and more useful. So impressed were the Romans with the Greeks' knowledge and skills that they often used Greek slaves as tutors for their children. The earliest specialist doctors and architects in Rome were typically Greeks. Romans modified Greek examples in engineering, architecture, geometry, physics, biology, and medicine. These included techniques in mining and grain mills, use of clocks and counters, aqueducts to bring fresh water to cities for baths and drinking, medical practices that made use of numerous plant remedies to ease suffering, and even the use and refinements of machines to delight and entertain crowds in amphitheaters.

Roman scientists, engineers, mathematicians, architects, and others left a rich legacy for future generations. Many Roman roads, aqueducts, bridges, mills, and sources of knowledge about STEM still exist today as evidence of their brilliant achievements.

CHAPTER ONE

THE SOLDIERS

April 21, 753 BCE, is the traditional date for the founding of Rome. In the midst of seven hills near the Tiber River in central Italy, Rome began as a farming village. Italy's first civilization, the Etruscans, conquered Rome around 600 BCE. Decades of strife followed, until the Romans overthrew the last of the Etruscan kings in 509 BCE and established the Roman Republic.

RISE AND FALL OF THE REPUBLIC

In the early days of the Roman Republic, Rome's influence covered about 400 square miles (1,036 sq km) and included a population of about 150,000. Two consuls served as generals and shared duties as heads of the state. Aristocrats, called patricians, voted in assemblies and occupied the Senate to debate and make the laws. Commoners, called plebeians, had no role in government, though they did have some protections under law. After plebeians staged an uprising in the late fifth century BCE, they obtained more power—until, by 367 BCE, the class had won the right to fill one of the consular positions.

Augustus Caesar was the first Roman emperor. Many consider him the greatest of all the emperors. Augustus, born Octavian, was adopted by his great-uncle Julius Caesar, whom he eventually succeeded as emperor.

Hardy and determined, the Romans expanded throughout the Italian peninsula, seizing lands and consolidating their power. They defeated Greek colonies in Italy by 275 BCE. Their next rival, from Carthage in North Africa, required more than a century of wars before Rome emerged victorious in 146 BCE. Steadily, Rome expanded through areas in Sicily, Macedonia, Greece, Asia Minor (today's Turkey), and Spain. They took many slaves from conquered people, who labored on farms and in cities alike. Educated Greek slaves taught many Roman children and served the elite as scribes, bookkeepers, and in other positions. Rome adopted the Greek gods—though they renamed them—philosophies, and scientific accomplishments.

All the while, Rome's armies continued to march through western Europe. General Julius Caesar defeated the Gauls in a series of wars from 58 to 51 BCE. Rome took over rule of the Gallic lands in modern-day northern Italy, France, Luxembourg, Belgium, and Switzerland. Caesar then marched back to Rome and took over the government. His reign lasted until 44 BCE, when his political enemies assassinated him. After a period of civil war, Octavian was appointed emperor and changed his name to Augustus in 31 BCE. The republic was no more.

THE MIGHTY EMPIRE

The Roman Republic may have ended, but the Roman Empire took its place. Called Imperial Rome, it lasted from 27 BCE to 476 CE. It began with Rome's first official emperor, Augustus. Beginning in 27 BCE, Rome entered a period of unprecedented peace and prosperity called the Pax Romana (Roman Peace), which lasted until 180 CE. During this time, Rome's army continued to expand the empire, which reached its fullest extent in 117 CE. From Britain in the north, it extended south to the countries of North Africa, including Egypt. From Portugal in the west to Mesopotamia in the east, the great empire dominated all of the Mediterranean Sea—and then some.

Though a brutally effective military force, Rome served as a relatively tolerant master. The centralized government allowed subject states to keep their own religion, so long as they worshipped the Roman emperor as well. The subjected peoples paid taxes to the empire and established coliseums, theaters, temples, and forums that duplicated those in the city of Rome.

Over a period of several hundred years, Rome became strained under the expenses of governing such a large empire (and maintaining the military). Taxes on the common people became oppressive and corruption plagued the government. The emperor's personal bodyguards assassinated their rulers frequently and appointed new ones, causing instability. Barbarian tribes from northern Europe and Huns from the east flooded into the empire. The empire divided into two: west and east. Eastern Rome—headquartered in today's Istanbul—grew stronger and wealthier, and eventually, in 476 CE, the last Roman emperor in the west was dethroned and the empire there collapsed.

THE SOLDIERS

Titled the *Mirror of Roman Magnificence*, this sixteenth-century engraving was produced during the Italian Renaissance, a time in which artists and thinkers idolized the accomplishments of ancient Rome.

THE ARMY'S IMPORTANCE

Rome's prosperity and great size would not have been possible without its conquering army. The army relied on cutting-edge (for the time) technology and incredible engineering feats, along with the common foot soldier—the legionnaire—to win the lands of Europe and beyond.

THE RIOTINTO MINES

Southwestern Spain has been the home of the Riotinto Mines, the largest-known ancient mining complex, for thousands of years. The Riotinto Mines produce silver and copper and have been in near-continuous operation since Roman times and well into the twenty-first century.

In 206 BCE, the Romans defeated and expelled the Carthaginians, a group of people from North Africa, who had held Spain since the sixth century BCE. With the help of skilled Roman engineers—and plenty of labor from convicts, slaves, and a few free men—the mine's peak production occurred from 70 to 180 CE.

The mine included shafts 450 feet (137 meters) deep, drainage channels—some of which remained in use until the twentieth century—and wooden water wheels.

The everyday life of the miners was brutal, and their lives were brief. For twelve hours a day, they labored in sulfur-choked air, swinging heavy hammers. Those with asthma or other breathing problems died quickly. Slaves and convicts wore chains at the neck and were fed a mixture of water and vinegar along with bread. When they died, they were thrown out with the garbage.

In Roman and pre-Roman times, the mine produced lead, a by-product of silver mining, arsenic, and other metals that poisoned the nearby Tinto and Odiel rivers—waterways still polluted with these substances today.

WEARABLE METALS

During the reign of Augustus (31 BCE–14 CE), the army consisted of about thirty legions of six thousand men each. The typical legionnaire wore a tunic covered by metal chain mail armor or plate metal armor made of bronze or iron. He fought with a short sword, javelin, and dagger. Officers wore a bronze chest piece called a cuirass. The metal used to forge these items came mostly from ore deposits, though Romans also mined tin, silver, and lead on a massive scale, mainly in Britain and Spain. Bronze, a mixture of copper and tin, creates a softer metal than iron. Iron nails pounded into the leather soles of the soldiers' sandals provided traction while marching and reduced the sole's wear.

Roman workers—often slaves—used stone mortars to crush ore. Millstones ground down the crushed ore into powder, then laborers washed it to concentrate it. To wash the ore, workers used instruments like sieves that could be moved from place to place, though some used permanent buildings called washeries. Through the use of aqueducts and huge tanks, they channeled water over the ore in a process called hushing. This caused the heavier, metallic pieces to settle in basins, while the dross was separated out. Metals were then heated and mixed in furnaces, after which they were fashioned into weapons or armor.

The water tanks were also used to release water so that it struck preheated rocks, causing them to fracture in a process called fire quenching. Breaking up rocks in this way assisted with locating ore and creating tunnels.

THE BIG GUNS

Roman artillery weapons, such as the ballista and the onager, provided the army with significant advantages over its enemies. These weapons were used in siege warfare against fortifications and towns and against armies on foot or horseback.

This nineteenth-century engraving shows a reconstructed Roman ballista. The fearsome ballista launched deadly stones and lead bolts up to 500 yards (457 m) with impressive accuracy.

The ballista came from Greece. Originally, it consisted of two wooden arms similar to those on a crossbow, bound by a twisted rope of horsehair or sinew, and attached inside a wooden frame shaped like a rectangle. A vertical stand with a slider allowed soldiers to load the device with heavy stones or lead bolts. It was armed by using a winch to pull the rope back. The device could stand 15 feet (4.5 m) high with arms 3 to 4 feet (1–2 m) long.

Roman engineers improved this Greek design by adding metal parts that made the ballista easier to assemble, lighter for transporting, more accurate, and significantly more powerful. In use from the third century BCE to the fourth century CE, the Roman ballista launched projectiles 450 to 500 yards (411–457 m). The missiles it launched easily pierced soldiers' body armor with lethal results. One variant of this weapon, the carroballista, sat on a cart to make it mobile. Each Roman legion had fifty-five carroballistae for use on the battlefield.

The onager, named after a wild donkey, catapulted missiles much heavier than those of the ballista, though only at a distance of 350 to 440 yards (320–402 m). With fewer moving parts than the ballista, the onager consisted of a large horizontal frame resting on the ground with a vertical frame on the front. Twisted animal hair or sinew ropes stretched a beam onto a horizontal position with winches to increase the torsion. A large stone rested in a cup or sling at the top. Release of the beam sent the stone into the air with violent force.

CHAPTER TWO

THE MEASURERS

Ancient peoples did not have access to a television set or computer screen for entertainment. They had another form of entertainment, one they could see every night: the sky. Without the glare of modern lights, the night sky is very dark, and stars and planets are much more easily visible. The movements of the planets and constellations fascinated ancient minds. Soon, people realized that they could tell time using the sky. The Romans were no exception.

STARSTRUCK

In modern times, astronomy is the study of the moon, sun, stars, and planets. Astrology, on the other hand, is a belief that the movement and positions of heavenly bodies influence the lives and fortunes of people on earth. Astronomy is a respected science in the modern world, but astrology is considered a supernatural belief unproven by science. However, in the ancient world, peoples often combined astronomy and astrology. The ancient Greeks studied astronomy, tracking the movements and qualities of celestial bodies, and developed theories about the way the heavens worked.

An Egyptian named Ptolemy (ca. 100–170 CE) was the most famous Roman star-seeker. He wrote thirteen volumes about astronomy called the *Almagest*. This work collected all the known astronomical knowledge from the Greeks and commented on it. Ptolemy is perhaps best known for his mistaken belief that Earth formed the center of the universe and that the sun, moon, and planets revolved around it. His claims became the foundation of much astronomical and scientific observation until the sixteenth century.

Greek astronomer Ptolemy wrote a thirteen-volume encyclopedia about astronomy that influenced the Western world until the sixteenth century.

Except for Ptolemy, the Romans had little interest in theories about astronomy. As in many STEM fields, they cared more about its practical use. A superstitious people, Romans embraced astrology because they believed it could tell their fortunes and help them predict the future. Fortune tellers would tell people's fates by examining the livers of animals that had been sacrificed to the gods, by observing the flights of birds in the sky, and other seemingly bizarre forms of divination. Astrologers cast people's horoscopes to predict what would happen to them in life by relying on the positions of the stars and other heavenly bodies on the date and time of their birth. Throughout the Roman Republic and Roman Empire, the belief in astrology flourished. However, it declined when Christianity became popular in the fourth century CE.

TIME MACHINES

The Romans recognized twenty-four hours in a day, just as modern societies do. They considered the day to be twelve hours long, and the night twelve hours long. The Romans adopted the Etruscan custom of having eight-day weeks. The eighth day was market day, when people did their shopping. Romans first used a lunar calendar, based on the cycle of the moon, from full to partial to disappearing for a few days. There is a problem with using a lunar calendar, though. A lunar calendar equals only 29 ½ days per month. This means that a year comes out to only 354 days. Because a year actually lasts 365 days, having eleven fewer days per year meant that before long, the months were out of sync with the seasons. Imagine if December occurred in the summer, for example.

Julius Caesar asked the Greek astronomer Sosigenes to solve the problem with the calendar. Sosigenes recommended that the Romans adopt the Egyptian calendar, based on the sun, not the moon. This solar calendar divided the year into twelve months with thirty or thirty-one days, except for February, which had twenty-eight days, except once every four years, when a leap year made it twenty-nine days long. The Julian calendar, as this new calendar came to be called, went into effect in 8 CE and remained in use until 1582, when it changed to the Gregorian calendar in use today.

SUNDIALS AND WATER CLOCKS

Ancient peoples recognized early on that they could tell the approximate time of the day by tracking the sun across the sky. The Babylonians and Egyptians invented the sundial, which was at first meant to measure which month it was. The Greeks constructed sundials that could measure more specific times of the day. These sundials were commonly stone with a concave surface. A pointer called a gnomon cast a shadow on the dial's surface, which was then interpreted to reflect the time of day. Romans copied this Greek design, but they took it one step further: they created portable sundials with a disc to be adjusted depending on the location. Roman towns had public sundials. Archaeologists have found thirty-five public sundials in Pompeii alone.

HOW STEM BUILT THE ROMAN EMPIRE

COUNTING ON AN ABACUS

The counting device known as the abacus existed in many ancient cultures, including Rome. Counting boards such as these had several vertical columns on which units, tens, or hundreds were inscribed, going from right to left. To perform addition, the column totals were carried to the left. The numbers could be marked by pegs, pebbles, counters, or in writing. One ancient Roman abacus displayed in Paris's National Library of France dates from the second century CE. The bronze plate has vertical grooves, in which beads are placed and slid up and down on numbered slots to indicate the count. The grooves are short (to hold just one bead of a value of five) or long (to hold four beads that represented one unit each). The board allows users to count fractions, ones, tens, hundreds, thousands, ten thousands, hundred thousands, and millions.

THE MEASURERS

Abacuses helped bankers, money changers, businesspeople, engineers, and others who needed to count as part of their work. These helpful mathematical tools remained in use for thousands of years until replaced by other calculators.

This replica of a Roman handheld abacus shows how ancient peoples calculated figures. The columns count up to very large sums on the left and very small sums on the right.

Another Roman timekeeping device was the water clock, an invention that probably evolved from the use of oil lamps. Lamps could be filled with oil to burn for a certain period of time. Water clocks released a set amount of water from one vessel to another, a process that took a set amount of time to complete. The Roman army used water clocks as a means of measuring work shifts. The Romans also used a type of Greek water clock called a *klepsydra*, which measured a time of about six minutes.

I, II, III, ET CETERA

The Roman mathematical system was so influential, it still exists in limited fashion in modern society. Today, Roman numerals appear in the ending credits of movies, the titles of Super Bowl games, and on clocks, to name a few places. The Roman numbering system used seven letters from the Latin alphabet to indicate values. So, I means 1; V means 5; X means 10; L means 50; C means 100; D means 500; and M means 1,000. A bar placed atop these letters meant that the numeral should be multiplied by 1,000 times its value.

Roman numerals evolved over time from the era of the Etruscans through the rule of Augustus. For instance, the number 100 was symbolized by a number of other markings until it eventually landed on the final symbol: C. Hand signals could also be used to signify specific numerals.

Numerals that appeared together were added if they were equal to or less than the first number. For

THE MEASURERS

ROMAN NUMERALS

1	I	11	XI	30	XXX
2	II	12	XII	40	XL
3	III	13	XIII	50	L
4	IV	14	XIV	60	LX
5	V	15	XV	70	LXX
6	VI	16	XVI	80	LXXX
7	VII	17	XVII	90	XC
8	VIII	18	XVIII	100	C
9	IX	19	XIX	500	D
10	X	20	XX	1000	M

This chart shows the values of Roman numerals from 1 to 1,000. These symbols still appear all over the world in various contexts.

example, II equaled 2 and VI equaled 6. Since this system might result in a long string of numerals for bigger numbers, a single numeral was never used more than three times in a row. In cases where a number would require a lot of repetition, a smaller numeral placed in front of a larger was subtracted from the total value. An example is the number 4, written as IV, or 9, written as IX. Multiplication and division with Roman numerals is complex, involving creating columns of numbers, striking out even numbers and identifying common factors. The Romans did not use the concept of zero.

CHAPTER THREE

THE TRAVELERS

The Roman system of roads spanned the length and breadth of the empire, for at least 50,000 miles (80,467 km), and it is one of Rome's greatest and longest-lasting accomplishments. Fragments of roads built during the empire survive in many countries today.

A well-designed and solidly constructed system of roads provided several benefits to the ancient empire: it encouraged trade between both close and far-flung destinations; it gave the military quick, reliable routes to travel to distant forts, deal with rebellions, conquer enemies, and send messages; it provided surfaces that allowed for the safe transport of food and supplies for the army; it stamped the countryside with a visible symbol of Rome's power and might; and it helped unify the different cultures of the empire into one.

ALL ROADS LEAD TO ROME

The most famous road was also the first, the Via Appia (or Appian Way) in Rome itself. Appius Claudius Caecus commissioned it in 312 BCE. Originally, it extended southeast from Rome to Capua for 132

This section of the Via Appia is in Minturnae, an archaeological site located between Rome and Naples. Rome's vast network of roads gave rise to the proverb "all roads lead to Rome."

miles (212 km). A later extension went to Brundisium (today's Brindisi) at the Adriatic coast. The Appian Way was so well constructed that it remains in use today.

By the second century BCE, four other roads radiated from Rome to the northwest (Via Aurelia), north (Via Flaminia), east (Via Valeria), and southeast (Via Latina), connecting the most important cities, such as Genoa and Capua, with Rome. Feeder roads branched off the major roads to connect smaller locales. As Rome conquered new territories, it built new roads through them until the network extended all the way across the empire.

The military constructed the most important roads, relying on the labor of the common legionnaire, each of whom carried a shovel as part of his standard equipment. Roman soldiers performed many construction duties. The most common was building the castra, or marching camp, built at the end of every traveling day. Constructed in a fixed layout, the castra was square or oblong, surrounded by a ditch and palisades, and included at least four gates. Such standardization not only promoted discipline (since it was accomplished immediately following each stop), but it also provided soldiers with a visual map of their camp in case they needed to flee an enemy's attack.

The Roman military employed specialists called *immunes* who were not required to participate in the foot soldiers' services. Instead, they performed skilled duties such as water engineering, blacksmithing, keeping sacrificial animals, piloting ships, making helmets, and more.

FEATURES OF ROMAN ROADS

Most Roman roads shared certain standard features. They were straight, not only because the shortest distance between two points is straight, but also because Roman wagons lacked fixed axels. This meant that the wagons could not be steered around curves— they could travel only in a straight line. Some roads, like those built in Pompeii, had ruts either built or worn into them to accommodate wagons and carts. When it was time to remove a vehicle from the road, workers had to use huge pry bars to inch it to the curb.

To build roads straight enough for the wagons that would use them, Roman engineers had to figure out how to tunnel through mountains, erect bridges, and build viaducts to connect hills and valleys. Such projects involved excavating land, transporting backfill, leveling the earth, diverting creeks, and draining large bodies of water.

A standard measurement for a Roman mile was 1,000 paces. This equals about .92 miles (1.48 km). Milestones date from 250 BCE onward; they recorded the names of consuls or other officials responsible for building and repairing roads and the distance from the city that erected it. Typically cylindrical, they were 6 feet (1.8 m) tall. A Roman odometer even existed for long-distance measurement.

AN ANCIENT ODOMETER?

Vitruvius, a Roman military engineer and architect, invented the Roman odometer around 15 BCE. His original design does not look like the one that appears in a modern car's dashboard, though. Instead, the odometer was a two-wheeled cart that resembled a wheelbarrow. A large gear with four hundred notches was attached to a wheel that sat in a box atop the vehicle. When the wheel made a complete revolution, it connected with a single-toothed hub mounted to the cart next to the wheel. When the wheel hub made a complete rotation, the tooth on it fit in the large notched gear, causing it to move one click. A complete revolution of this toothed wheel activated a metal hand that fit into the teeth of another wheel—resembling a saw blade—in which sat a number of small stones. One stone then fell into an indentation in the top of the box to signify that four hundred revolutions had passed. Four hundred revolutions equalled one Roman mile. The process could then repeat itself and the distance traveled by this odometer could be calculated.

Some have speculated that the odometer was not actually built by Vitruvius originally, but by the legendary Greek mathematician and inventor Archimedes much earlier.

A celebrated Roman writer, architect, and inventor, Vitruvius lived and worked in the first century BCE. He may have been the creator of the Roman odometer.

CONSTRUCTING THE ROAD

The width of major Roman roads was 13.8 feet (4.2 m), or enough space for two wheeled vehicles to pass one another. A packed-gravel pedestrian path ran next to each side of the road, varying in width from 3 to 10 feet (1–3 m). Upright slabs of stone formed the curb. Every 10 to 16 feet (3–5 m), a higher curb block allowed for people to mount their pack animals or horses. Some busy stretches of roads had pull-out areas to allow traffic to exit the road temporarily.

The road itself was built so that it was slightly higher in the center, which allowed rainfall to run to the side, where drains and drainage ditches were often dug. This prolonged the lifespan of the road's materials.

The first step in laying a new roadway involved land surveying. The most important surveying instrument was the *groma*, which assisted in the creation of straight lines, rectangles, and squares. It consisted of a wooden pole with a cross on the top. Plumb lines, or small weights on strings, hung from each of the cross's arms. The groma was planted in the ground and twisted until the surveyor could sight along two of the strings to the starting location. Once that was determined, the surveyor walked around the groma to sight the other direction to a second point. If the strings did not line up to this second point, it was moved until it did so. This process was repeated from the second point so that the plumb lines lined up with the starting point—indicating a straight line.

Local building materials were used during road construction. First, a trench was dug and a foundation of gravel or crushed bricks, clay—or in marshy areas, wooden piles—was laid inside. A layer of finer gravel went on top of this, to be covered with stone blocks or slabs. A gravel surface mixed with lime lay over the top of it. In more opulent or wealthy areas, the top finish was made of basalt or limestone pavers, cobblestones, or blocks of volcanic tufa.

TUNNELS

Rome adapted ancient Persian techniques when it became necessary to dig tunnels through mountains so that roads could pass through them. The three most important tunnels were built in the first century BCE and ran between 766 and 1,094 feet (700–1,000 m) in length.

The process of digging tunnels often involved excavating earth from both ends at the same time, a practice called counterexcavation. Precise calculations were needed to be sure that both ends of the tunnels met. As work proceeded, shafts were sometimes drilled into tunnels from above to make sure both tunnels would meet. Shafts also allowed workers to speed up excavation by approaching the work from another angle.

CHAPTER FOUR

THE BUILDERS

The ancient Romans built massive aqueducts, long-standing bridges, and awe-inspiring temples. The main engineering triumphs of Roman engineering and architecture came from their use of arches and domes, advances that left a lasting impact on architectural history.

BRICKS AND ARCHES

Stones made up the earliest Roman building material. One stone was placed atop another and no mortar was used to hold them together. By the sixth century BCE, the Romans had developed techniques to cut blocks of tufa and limestone for use in walls. Sun-dried mud bricks were made by mixing water with sand, silt, or clay and allowing them to dry in the sun. By Augustus's reign in the first century BCE, Romans had adapted a Greek technique for making fired-clay bricks. The bricks were mixed, formed, dried, and placed into an oven about 1,800 degrees Fahrenheit (about 1,000 degrees Celsius). Roman bricks were a different size from modern bricks, measuring more than 3 feet (90 centimeters) in length. Fired bricks were much stronger than sun-dried bricks and became the primary building material for monuments, baths, and other construction projects.

CROSSING OVER

In addition to advances in brickmaking, Rome's mastery of the arch allowed for longer roads, larger buildings, and sounder aqueducts. Romans did not invent arches—they were used in both Egypt and Greece, although in smaller buildings—but the Romans were the first people to realize how important arches were for building bridges.

The Romans, assisted by the arch, were master bridge builders. Arches are constructed by building

The Arch of Hadrian dates from 129 BCE, when it was built in Gerasa, Jordan, to honor Emperor Hadrian's visit. The arch is an iconic hallmark of Roman construction.

a wooden frame that allows wedge-shaped blocks to be held in place as they are built into a semicircular shape. The keystone is the last block put in place in the exact center of the arch. The weight of the keystone presses down throughout the rest of the arch, allowing the entire construction to stand firmly when the wooden frame is removed.

Arches allowed Romans to build stronger, longer bridges. One famous example is the bridge spanning the Tagus River in the town of Alcántara, Spain. Built in 105 to 106 CE in honor of the Roman emperor Trajan, it runs 636 feet (194 m) long. Though the arches in the bridge are different sizes, some span about 98 feet (30 m), with huge arch stones holding it all together. The arch stones were fitted so closely that the joints required no mortar. The bridge still stands today, after nearly two thousand years; hundreds of other Roman bridges still stand as well.

The first stone bridges made use of iron clamps to hold stone blocks together, but concrete was used by the middle of the second century BCE. Often, bridges included a concrete core faced with stone blocks. Strong piers held the load of the structure at each end.

Roman use of concrete is famed. Though concrete originated in ancient Babylon, Romans began to use it by 200 BCE. Roman concrete used volcanic ash, lime or gypsum, and stone mixed with water to create a chemical reaction that made the substance strong and durable. Romans developed concrete that could harden underwater by the mid–first century CE, an essential quality for constructing Rome's bridges and aqueducts.

THE WATER OF LIFE

Rome, like all civilizations, had a great need for water. Water was used for drinking, nourishing crops, flushing waste in sewers, filling public baths, supplying fountains, and even powering mills and other machines. All this water came from one of Rome's most enduring and innovative STEM legacies: the aqueduct.

An aqueduct is a bridge-like channel that is supported by tall, arched structures. Aqueducts were known in the ancient Near East, but the Romans used them on a scale never before seen between 312 BCE and 455 CE. Aqueducts up to 57 miles (92 km) long brought water across the land to cities using a series of pipes, tunnels, and conduits. The basic principle behind aqueducts was gravity, not pumps. Aqueducts had a slight downward slant so that water would naturally flow to its destination. Arcades—or arched, covered passageways—allowed water to travel over valleys and low terrain, and tunnels sent water underground. Settling tanks at set intervals helped to regulate the supply of water. Aqueducts also employed a number of valves, siphons, and mesh filters to improve the transportation process.

Eleven major aqueducts supplied water to the city of Rome. The longest of these aqueducts is the Aqua Marcia, built in 144 BCE. Fifty-seven miles (92 km) long, it ran through an underground trench for 50 miles (80 km), and flowed along an arcade aboveground for the final 7 miles (11 km). Reservoirs

HOW STEM BUILT THE ROMAN EMPIRE

This aqueduct from the south of Spain demonstrates Rome's considerable skill in construction. The fact that Roman aqueducts and other buildings still stand is also a testament to their ingenuity.

contained the water at the destination. Pipes extended from these reservoirs to public fountains, sewers, and directly to the homes of those few citizens who could afford it.

Aqueduct water channels were made from concrete, stone, brick, or rubble. Pipes were made from clay or lead. Complete or partial aqueducts survive today from the ancient Roman Empire.

ON ARCHITECTURE

Vitruvius (c. 90–20 BCE) was a Roman architect and military engineer who authored *De Architectura* (*On Architecture*), a book about feats of ancient engineering and architecture. Much of what is known about Roman architecture today comes from this work. Based on Greek architecture and his own experience, *De Architectura* contains ten sections on building materials, construction techniques, public and private structures, hydraulics, clocks, astronomy, firearms, missiles, the use of sundials, and water clocks. The broader STEM topics it covers are geometry, mathematics, astrology, science, medicine, meteorology, and philosophy. Though some of these subjects are not considered relevant to architecture today, they were vitally important during Roman times.

(continued on the next page)

(continued from the previous page)

Vitruvius's treatise also contained famous stories, like that of the moment the Greek mathematician Archimedes discovered the law of buoyancy. However, many of his claims are considered mythical or legendary, rather than historically accurate. Still, the information is useful in understanding the way the Romans viewed their earlier neighboring culture.

TEMPLES TO THE GODS

The widespread use of another iconic Roman building feature—the dome—came about around 100 CE. When Roman builders expanded an arch into a circle, they realized that they could create a three-dimensional shape with increased strength; this was eventually identified as a dome.

The dome shape exerts thrust, or pressure, around its circumference. Heavy walls had to support the dome. One well-known example of the dome occurs in the Pantheon, a temple to all the Roman gods, begun in 27 BCE and then rebuilt under Emperor Hadrian (76–138 CE) in Rome. A circular building made of concrete and brick, it includes a massive concrete dome decorated with coffers, or recessed square shapes that helped reduce the weight of the dome. At the topmost point of the dome, there is an oculus, or hole. Since the dome is weakest at this point, the builders did not use stone or concrete at all at this location. The oculus was also used to let in light. It is designed in such a way that every year at noon on Rome's traditional birth date—April 21—light shines through the oculus and out the door.

The Pantheon's dome remained the largest in the world until modern times, measuring 142 feet (43 m) high and having roughly the same diameter. Arches and piers were constructed to strengthen the dome. The Pantheon exists today as a major tourist attraction in Rome, and the building still boasts the largest unreinforced stone dome in the world.

Oculus means "eye" in Latin. The oculus in Rome's Pantheon lets in light, directs the viewers' attention to the heavens above, and lightens the weight of the dome.

CHAPTER FIVE

THE FARMERS

Despite the glory of the city of Rome and other large cities across the empire, such as Alexandria, London, and Cologne, most people lived in farms scattered across the countryside. The majority of crops they grew were grains, grapes, and olives. Though these people did not directly contribute to flashy Roman construction projects, the advancements they made in the STEM field of agriculture were no less important.

FARMING TECHNIQUES AND TECHNOLOGY

The plains around Rome were rich in volcanic soil that provided the earth with beneficial potash and phosphate. However, the soil on the surface was thin and easily overfarmed. The earliest farmers used crude wooden plows pulled by horses or oxen to scratch thin furrows in the land, in which they planted seeds. Repeated plowing, straight and at an angle (cross-plowing) caused the soil to become nearly dustlike. This consistency helped the seeds extract maximum nutrients from the ground. Over time,

THE FARMERS

Romans started using iron cutting blades on their plows because of the material's increased strength and durability, especially when plowing through tough soils like clay.

Romans used two-pronged mattocks (hand tools like pickaxes), shovels, and hoes in farming. They did not have the wheelbarrow, but instead used carts, baskets, or sacks to carry their loads. Around the first century CE, wheeled plows came into use.

This intricate mosaic from Algeria shows oxen pulling plows and farmers urging them on with whips. Olive trees serve as background decorations.

The shape of the land—its topography—could cause water troubles for farmers. To keep water from accumulating in low-lying areas, where it could breed insects and drown crops, they dug drainage ditches. These ditches channeled water runoff away from fields. Flash floods sometimes carved deep gullies in fields, making drainage ditches even more important. On land that sloped, terraces were dug to keep water from eroding the soil. Vines, olives, and fruit trees often grew there.

Fertilizer—in the form of manure from horses, sheep, and cows—enriched the soil. Compost pits held food waste, ashes, straw, leaves, and manure. It was stirred from time to time as it broke down to become a rich source of fertilizer.

Roman farmers learned that greenhouses could increase their success rate with growing some plants. They were proficient with pruning trees to shape them and increase the amount of fruit they grew. Grafting, or joining two or more plants together so that they grow as one, is a modern technique that the Romans used as well. Roman farmers even experimented with crossing pumpkins and apples to make a hybrid fruit.

Romans preserved food so that it would last longer through various means, including storing it in honey, pickling it in vinegar or brine, drying, smoking, and salting it. They raised game animals like deer, boar, and rabbits for food by containing them in large, supervised forested areas. They built large tanks to contain fish. Farmers also raised pigs, goats, cows, sheep, and poultry.

THE WRITER COLUMELLA

Roman writer Columella lived during the first century CE. Born in Spain, Columella learned about farming techniques while young. He served as a soldier and tribune during his lifetime but found farming in Italy to be a more satisfying line of work. He advocated a simple farming life in his writings, which included a twelve-book series called *De Re Rustica* (*On Rural Affairs*). The twelve books covered agriculture from many angles, including how to run an estate, grow grains, cultivate vineyards and trees, breed cattle and poultry, and keep bees. Columella also wrote about farm calendars, climate, soil types, and techniques that yielded the greatest amount of crops. Unfortunately, little of *De Re Rustica* has survived, but it is still an important work in the history of Roman STEM.

Columella wrote about farming to inspire readers to take up a simple, rural lifestyle. Two of his books were translated into English as *Of Husbandry* in 1745.

MILLS, PRESSES, AND MORE

Grain has to be ground down in order to produce flour. To manufacture their flour, Romans used mills. No matter the size, all mills have two main parts: an upper millstone that moves and a lower, larger millstone that does not move. The grain is placed between the two and ground.

Early hand mills were made by placing a wooden handle in the upper stone. People—typically slaves—then moved the stone to grind the grain. Larger forms of this mill type were powered by mules, donkeys, or horses that were blindfolded and tied to the handle so that they could turn the mill as they walked in a circle.

Animals or people could generate limited amounts of power, however, and even slave labor was relatively expensive over long periods of time. The use of water provided much more, and cheaper, power in the form of waterwheels. The Greeks used a wooden wheel with six to eight water scoops attached to a shaft. When the wheel was placed sideways in a quick-flowing stream, the scoops caught the water and turned, causing the shaft to turn as well. The shaft was connected to the millstone, which then turned to grind the grain. However, this form of waterwheel, just like human and animal power, had limits.

The Romans adapted the Greek waterwheel design and improved it. They turned the wheel upright, or vertical, to make two types of waterwheels. One type—called an undershot wheel—had blades pushed by the current of moving water. The blades were attached to gears that turned to run the mill. The overshot wheel was more powerful and efficient than previous versions. It made use of water falling from a chute over the top of the wheel. The weight of the water, brought down naturally by gravity, caused the overshot wheel to turn more quickly. The water from the chute was often supplied by a nearby aqueduct or tall reservoir.

A gigantic example of the overshot water mill remains in Barbegal, France, near the town of Arles. Built in the fourth century CE, it generated power through a system of sixteen waterwheels. The wheels were placed in two parallel rows of eight. Water ran downhill from one wheel to the next, until it ran into a drain at the bottom of the hill. Modern scholars calculate that the mills produced about 9,900 pounds (4,491 kilograms) of flour per day, enough to feed the population of Arles, which consisted of about 12,500 people.

PRESSES AND OLIVES

Olive trees are the perfect plant for warm Mediterranean climates like those in Rome, Greece, and Spain. The tree requires little maintenance, lives a long time, and resists drought. Olives can be eaten when brined, or soaked in salt water, and olive oil is a useful food product. Olives must be pressed in order to extract the oil. This could be done by foot, by using a mortar and pestle, or using a stone roller. An olive press is needed for large production. The press crushes olives with a round stone that consists of a big stone bowl to hold the olives that were then

This old mosaic from France shows men using an olive press to extract valuable olive oil. Olives and olive oil continue to be heavily exported from Mediterranean countries today.

pressed between two concave blocks attached to a beam and iron pivot. This device fit into a central post. Often, mules would turn the press. Other methods included beam and winch presses—which featured a beam attached to a wall and a stone winched down to crush the olives—and screw presses that drove a ram up and down by a screw.

CHAPTER SIX

THE HELPERS

Ancient Rome made a number of advances in medicine and public health, helping soldiers and citizens alike live longer, healthier lives. Roman contributions in this area of STEM remained prevalent for centuries.

MEDICINE MAKES A MARK

According to the scientist and writer Pliny the Elder (23–79 CE), traditional Roman medicine was performed by the head of each family, a practice that had varying benefits or none at all. This was in contrast to the practices of the earlier Greeks, some of whom had made medical practice their vocation.

As with nearly all of the Romans' STEM accomplishments, they drew upon the work of the Greeks. One of the first Greek physicians to practice in Rome, Archagathus of Sparta, arrived in 219 BCE. A specialist in treating soldiers' battle wounds, he also gained fame for improving skin conditions. He grew so famous that Rome granted him citizenship, but his practice of cutting into bodies and removing limbs was not received well by Roman scientists.

HOW STEM BUILT THE ROMAN EMPIRE

Pliny the Elder wrote an encyclopedia titled *Natural History*. He died from inhaling poisonous gases in 79 CE when he watched the eruption of Mt. Vesuvius from a nearby harbor.

The Greek physician Asclepiades of Bithynia (124–40 BCE) brought considerable medical knowledge to Rome. Instead of the popular therapies of medications and bloodletting, he prescribed music, sunlight, walks, massages, hot and cold baths, and special diets for his patients. At the time, people diagnosed with mental health problems were often cruelly confined to dark rooms. Asclepiades improved their treatment by releasing them and using exercise, sedatives, and musical therapy. He believed that throughout the body, there were empty spaces—called pores—in which atoms existed, and it is thought that Asclepiades used tracheotomy as an emergency therapy for breathing difficulties.

FIGHTING AND HEALING

The army and the wealthier private estates had their own doctors. Battlefield injuries provided surgeons with practical experience. These physicians carried a tool kit that included arrow extractors, scalpels, forceps, and catheters. They relieved pain with opium from poppy juice and used acid vinegar to clean wounds. From the first century CE onward, each army unit had its own hospitals. Special hospitals also existed for the treatment of slaves.

Public hospitals did not come into existence until the Christian era. Originally, hospices sheltered pilgrims and messengers for the clergy. The first hospital in the city of Rome was built by the wealthy widow Fabiola in the fourth century CE. The modern hospital has its roots in these early institutions.

LIVING IN GLASS HOUSES

Romans made wide use of glass in everyday life. Glass bottles held medicine, while glass plates and cups were used to serve food and drinks. Higher-quality glass was used in mosaics, mirrors, jewelry, sculpture, magnifying glasses, and even windows. Glass ground into powder was even used as medicine and toothpaste.

The Egyptians and Mesopotamians created glass as early as the fifteenth century BCE, and it later made its way to the Etruscans, from whom the Romans inherited it. During the Roman Republic, glass was mainly used to contain expensive oils, perfumes, and medicines. By the time of the Roman Empire in mid-first century CE, Roman glassmaking had exploded in popularity and was used to make glass panes for homes and for greenhouses.

Glass consists of a mix of limestone, silica, and soda ash heated to a liquid consistency, formed into the desired shape, and then allowed to cool. Early methods of creating glass involved casting it into molds, and different colors were added for decoration. The vessels this method produced were larger, such as jugs, bowls, and plates. By the first century BCE, glass blowing was in use.

To blow glass, the ingredients are heated to liquid form at the top of a hollow iron rod. Once hot enough, the artisan blows through this tube to inflate and manipulate the liquid. This method created thinner, higher-quality glass that could be produced quicker and more cheaply.

Many Roman glass objects rival the finest, most expensive art produced during antiquity, incorporating a variety of shapes, colors, and designs with a gemlike appearance.

FAMOUS PHYSICIANS AND THEIR PLANTS

Roman physicians made use of a variety of herbs and plants for all sorts of ailments. Pliny the Elder advocated unwashed wool soaked in fat and the herb rue for swellings and bruises and dipped wool in oil to solve uterine problems. Statesman and writer Cato the Elder (234–149 BCE) believed that eating cabbage fended off many diseases and that drinking the urine of cabbage eaters provided health benefits. Celsus (25 BCE–50 CE) authored an eight-volume encyclopedia on medicine that covered the beneficial effects of diet, therapy, and surgery. He identified about 250 medicinal plants, including the spices pepper, cinnamon, and cardamom.

The most prominent writer about plant medicine was Dioscorides (40–c. 90 CE), an army surgeon for the Roman emperor Nero. He wrote about more than 650 plant remedies, including garlic, onions, chamomile, willow, sage, coriander, and parsley. Modern scientists have identified antibiotic properties in garlic and onions, and willow bark is the source

This page from one of Dioscorides's books gives information about yarrow, foxglove, and primrose plants. For centuries, books from ancient Rome provided Europeans with knowledge about the world around them.

of pain-killing chemicals with effects similar to those of aspirin.

Galen (129–ca. 199 CE) is the greatest physician from the Roman Empire. Born a Greek from Asia Minor, he learned to be a doctor by treating the wounds of gladiators. He advocated the use of dissection to learn about human anatomy. His observations about the kidney, bladder, liver, and heart were accurate. Traveling to Rome in 161 CE, he became the well-respected personal physician to the emperor Marcus Aurelius and his son Commodus. A prolific writer, he penned more than four hundred texts on medicine, philosophy, and drama. His writings influenced European history for more than a thousand years, making him one of the most important sources of Roman knowledge.

SPECIALTIES AND DIVINITIES

One of the enduring themes in Rome's STEM history is the dedication to real, practical applications for knowledge. To that end, many Roman physicians and scientists developed or practiced highly specialized medicine.

Ancient Roman law dictated that when any woman died during labor, the child must be cut from her. This law may be the source of what is today called a Cesarean section, which involves cutting a newborn from its mother's womb. In Roman times, this was done for two reasons: first, by tradition, pregnant women could not be buried with a child in the womb; if a woman died while in labor, the child was removed to allow her to be buried. Second, doctors realized that such a cutting procedure could help save a baby's life. In modern times, this is a procedure typically performed on living women who cannot safely deliver a child vaginally.

Medical instruments used to treat cataracts dating from ancient Rome have been found. The procedure they used, called couching, would have involved pushing the clouded lens to the bottom or rear of the eyeball, which may have resulted in an increase in the light received—but it also frequently resulted in total blindness.

Because Roman medicine was limited by a lack of understanding and knowledge, such as of the existence of germs, much of it was ineffective. Because of this, many Romans resorted to magical spells and divination to look for a cure to what ailed them.

ATTENDING TO PUBLIC HEALTH

Ancient Romans put a premium on cleanliness. They had discovered that not only did such practices make one more pleasant to be around, but they also had distinct health benefits.

BATHING BEAUTIES

Baths served as more than just places to get clean. They were a central part of Roman life where rich and poor alike met and relaxed. Soldiers on campaign even built their own baths in the field. Baths varied in size depending on the town, but they were commonly located near the forum—the center of urban life.

THE HELPERS

Large cities could support monumental baths that accommodated thousands of visitors each day. Such baths were massive complexes with colonnades, wide arches, domes, mosaic floors, marble walls, and beautiful statues. The large, airy space needed for baths drove architects to make use of the dome as never before.

Baths had a number of rooms designed to be visited in this order: changing rooms, exercise rooms, open-air swimming pools, sweating rooms, hot rooms with hot water pools, warm rooms, cool rooms, and rooms for massage. Baths prior to the first century BCE used braziers for heat, but afterward they used under-floor heating fuelled by furnaces. Originally a

These ruins show the interior of a Roman bath. Note the holes and chambers that served to deliver heated air and water to various rooms.

Greek idea, indoor heating systems called hypocausts were made bigger and better by the Romans. In a hypocaust system, furnace fires were located under a raised floor. Hollow tubes and specially designed bricks then conveyed heated air to appropriate rooms.

The Baths of Caracalla were among the most famous in Rome. Construction on the building started in 206 CE under the supervision of Emperor Septimius Severus; his son Caracalla completed them in 216. They could hold up to 1,600 bathers at a time. The bath chambers were made from vaults to cover an area of 750 by 380 feet (230 by 115 m). The structure included numerous rooms and a garden where exercise and games were held. The Baths of Caracalla continued to be used through the sixth century.

ROME'S GREAT SEWER

In Rome, sewers and drains were flushed out by aqueduct water. Even if they did not know the exact reasons, Roman engineers and scientists knew that it was unhygienic to leave waste to rot in pipes. One of Rome's oldest monuments is known as the Cloaca Maxima (Great Drain). It was first used as an open, stone-lined channel in the sixth century BCE, making it one of the world's oldest sewer systems. In the third century BCE, it was enclosed by a barrel-shaped vault. Its main function was to carry stormwater from the forum into the Tiber River. Under imperial Roman rule, it was connected to latrines and large public baths.

CHAPTER SEVEN

THE IMMORTALS

From amphitheaters and ingenious inventions to long-lived scientific texts, much about ancient Rome influenced the world throughout the centuries and still remains accessible—and sometimes even usable—today.

A FAMOUS AMPHITHEATER

Rome produced many iconic structures, and among the most common were amphitheaters. It was at these venues—oval or round stadium-like structures—that Romans watched gladiatorial games, contests between men and wild animals, public executions, and mock naval battles. The largest amphitheaters could seat tens of thousands of spectators. In all, more than two hundred amphitheaters were built across the Roman Empire, often over the ruins of older Greek stadiums, which served a similar purpose.

The biggest and best-known amphitheater is Rome's Colosseum, also called the Flavian Amphitheater, which opened in 80 CE. This structure is made from stone and concrete in an oval shape. While most amphitheaters were built in a circular shape, the Colosseum was shaped like an oval to provide all the

Rome's Colosseum is one of the city's most famous buildings. The holes visible around the arches were once filled with iron supports that were stolen and reused throughout the centuries.

spectators with a good view. The structure is ringed by rows of arches on each floor, and it employed a number of barrel vaults—or series of arches joined together—to make a roof or ceiling, along with groin vaults, which were formed by placing two barrel vaults at right angles. These vaults measured 620 by 513 feet (189 by 156 m). With eighty entrances, the

Colosseum stands 157 feet (48 m) tall and it could hold about fifty thousand spectators.

The spectacles held at the Colosseum included gladiator fights, fights between men and wild animals, public executions, and other forms of entertainment. The Colosseum's arches and seats still stand, but so does something else: its underground chambers, called the hypogeum. In its day, a wooden floor sat atop the hypogeum. The hypogeum contains a series of walls, chambers, and stairways. Over time, after Rome's fall in 476 CE, the wooden floor eroded and knowledge of the purpose behind the walls, chambers, grooves, cuts, masonry, slots, and abrasions located in the hypogeum was lost.

Ancient sources talked about how gladiators fought wild animals like bears, wolves, and lions that rose out of hidden trap doors in the Colosseum's floor. For instance, Emperor Titus held a number of games to celebrate the opening of the Colosseum, during which thousands of animals were killed.

Modern archaeologists have figured out that many of the tracks and channels under the floor contained large capstans, or devices designed to move heavy weights by use of cables, ropes, or chains. The capstans were used to lift and lower platforms or cages containing wild animals.

By building scale models, archaeologists determined that the hypogeum contained sixty capstans, each of which stood two stories tall and required the labor of four slaves to operate. Animal cages use forty of these capstans, while twenty raised and lowered large platforms that delivered scenery to the floor above. Twenty-eight smaller platforms for additional scenery and props operated with a series of ramps, hoists, counterweights, and cables.

Archaeologists have also identified traces of runoff canals. These were used beginning in 46 BCE to drain the Colosseum after it was flooded by a nearby aqueduct. Why was it flooded? Historians believe the Romans used the Colosseum to reenact naval battles on an artificial lake, with warships and all.

SPECIAL EFFECTS

Chariot races and other outdoor entertainment was often accompanied by the loud, penetrating sounds of a *hydraulis*, or a mechanical water organ. A Greek named Ctesibius of Alexandria invented it in the third century BCE and brought it to the Roman world. The organ consisted of a set of pipes on a wind chest attached to a cone-shaped wind reservoir. One or two pumps supplied the reservoir with air. An outer container—half-filled with water—sat outside the wind reservoir. Air pressure was adjusted when the player operated keys or sliders that let air into the pipes, creating the loud tones.

In the first century CE, Heron of Alexandria invented a thunder machine for use in Roman theatre. By activating a lever, a number of brass balls were released to bounce down a series of shelves before falling onto a tin sheet and creating deep bass tones that sounded like thunder. Though both of these sound pioneers' inventions seem to be rudimentary by modern standards, they helped immerse the Romans

THE IMMORTALS

This example of an ancient Roman theater comes from a town on Libya's coast. The semicircle design of seating is often employed in modern theaters to effectively carry sound to spectators.

in the events they were viewing—and, with the aid of these devices, hearing.

ROME'S ENDURING LEGACY

Throughout much of modern Western civilization, the impact of Roman STEM innovation can still be seen. Roman architectural features, such as arches, vaults, and concrete, continue to influence construction today.

THE CHRONICLERS

A number of Roman authors passed STEM ideas and concepts on to later generations. While many writings are lost to scholars, a number of influential texts remain.

Cato the Elder's *De Agricultura* (*On Agriculture*) not only advised readers about how to run an estate, but also how to increase the production of wine and oil and how to implement solutions to crop diseases.

Varro (116–27 BCE) is known as one of Rome's greatest scholars. He wrote hundreds of texts on a wide variety of subjects, including law, education, astronomy, geography, mathematics, and biology, along with a number of letters, speeches, and poems. Though the majority of his work is lost, *Res Rusticae* (*Farm Topics*) survives in a complete form. This guide gives practical advice about general agriculture and animal husbandry. Later authors, such as Vitruvius and Pliny the Elder, drew heavily on his work when writing their own.

Pliny the Elder completed *Natural History* in 77 CE. The first science-based encyclopedia, it covered nearly every topic known to science at the time. However, it had a number of inaccuracies since Pliny did not distinguish between speculation, opinion, and fact. Nevertheless, the thirty-seven volumes of *Natural History* served as much of Europe's primary source in

THE IMMORTALS

scientific knowledge until the Middle Ages. The subjects covered included astronomy, zoology, botany, medicine, and geology.

Pliny's curiosity ended up being the death of him: he was killed by the poisonous gases he inhaled while trying to observe Mt. Vesuvius's eruption in 79 CE.

Columns, temples, and amphitheaters still stand as ruins—and reminders of the ancient civilization's accomplishments. Some structures, such as aqueducts, bridges, and roads, still remain in use.

Romans excelled in most STEM fields, but particularly so in engineering. They understood the principles of physics well enough to develop massive aqueducts. Their waterwheels powered immense mills and enabled them to drill through stone in mines.

Practical farmers at heart, Romans also demonstrated an extreme level of agricultural advancement in their practices of crop rotation, fertilizer, and grafting, and they used large-scale mills to create sophisticated factories that would not be replicated until the modern age. They developed medicine to a relatively high degree and applied a similar effort to their sports and entertainment.

While all these achievements are impressive, perhaps the greatest of all is the knowledge Romans passed down about medicine, the natural world, construction, and principles of science, which has been great enough to last for generations.

TIMELINE

April 21, 753 BCE   The traditional date for the founding of Rome.

600 BCE   The Etruscans conquer Rome and establish a monarchy.

509 BCE   The Romans overthrow Etruscan kings and establish a republic.

300 BCE–400 CE   The ballista is refined and used by the Romans.

312 BCE   The Appian Way is commissioned.

206 BCE   The Romans expel the Carthaginians from Spain.

200 BCE   Concrete comes into use around Rome.

146 BCE   Rome emerges victorious against its long-time North African rival, Carthage.

100 BCE   The dome comes into widespread use.

58–51 BCE   Julius Caesar fights a series of wars in Gaul.

44 BCE   Julius Caesar is murdered.

31 BCE   The Roman Republic ends, and the Roman Empire begins with the rule of Augustus.

TIMELINE

8 CE   The Julian calendar goes into effect across Rome.

80 CE   The Colosseum is opened in Rome.

117 CE   The Roman Empire reaches its greatest extent.

216 CE   The Baths of Caracalla are opened.

476 CE   The Roman Empire in the West falls.

GLOSSARY

arcade A row of arches with supporting colonnades.
backfill To refill a hole with the earth that was dug out of it.
colonnade A row of columns used to support the base of a roof.
conduit A channel to carry water.
constellation A pattern formed by stars.
dissection The act of cutting open a dead body to learn about its internal organs.
divination Seeking knowledge about the future by the use of omens or through other supernatural means.
dross Waste that forms on molten metal.
forum The center of public business in a Roman town.
horoscopes Forecasts about a person's future based on the positions of the stars and planets at their birth.
husbandry The cultivation, care, and breeding of animals and crops.
hydraulics The movement of liquids through channels and pipes.
mortar A mixture of lime and cement spread between bricks or stones to hold them together after it hardens.
mosaic A piece of art made from small pieces of tile, stone, or glass set on a hard surface and arranged to make pictures or patterns.
oculus A round opening in a dome.
pier A support for a bridge or building.
plumb line A cord with a weight on one end that is used to determine a vertical line.

GLOSSARY

reservoir An artificial lake that stores water.
settling tank A tank designed to hold water until the particles within it settle to the bottom.
sinew A piece of tough tissue that attaches muscle to bone or bone to bone.
torsion The act of twisting a material to create force.
tufa A porous rock made of calcium carbonate that is formed by water droplets.
wooden piles Large, long wooden columns driven into the ground to serve as support for a structure.

FOR MORE INFORMATION

The Getty Villa
17985 Pacific Coast Highway
Pacific Palisades, CA 90272
(310) 440-7300
Website: http://www.getty.edu/visit/villa
Facebook and Instagram: @gettyvilla
Twitter: @GettyMuseum
Designed like an ancient Roman villa—but constructed in Malibu, California, within sight of the ocean—this museum location focuses on ancient art through Roman times.

Metropolitan Museum of Art
1000 Fifth Avenue
New York, NY 10028
(212) 535-7710
Website: https://www.metmuseum.org
Facebook, Twitter, and Instagram: @metmuseum
The museum contains and displays more than 1,500 artifacts and artworks from ancient Rome.

Montreal Museum of Fine Arts
1380, Sherbrooke Street O
Montreal, QC H3G 1J5
Canada
(514) 285-2000
Website: https://www.mbam.qc.ca/en
Facebook, Twitter, and Instagram: @mbamtl
The museum houses hundreds of objects from the ancient Mediterranean, including Greek, Roman, Egyptian, and Middle Eastern art and household items.

FOR MORE INFORMATION

Oriental Institute
University of Chicago
1155 E 58th Street
Chicago, IL 60637
Website: https://oi.uchicago.edu
Facebook: @OrientalInstitute
Twitter: @orientalinst
The institute features an extensive collection of artifacts, writings, and artwork from ancient cultures, including Rome, Greece, Mesopotamia, and more.

Penn Museum
3260 South Street
Philadelphia, PA 19104
(215) 898-4000
Website: https://www.penn.museum
Facebook, Twitter, and Instagram: @pennmuseum
This museum's Rome gallery boasts impressive sculptures and statues dating back to ancient Rome.

Royal Ontario Museum
100 Queen's Park
Toronto, ON M5S 2C6
Canada
(416) 586-8000
Website: https://www.rom.on.ca/en
Facebook: @royalontariomuseum
Twitter and Instagram: @ROMtoronto
The Eaton Gallery of Rome features more than five hundred artifacts that paint an intriguing picture of daily life in ancient Rome.

FOR FURTHER READING

Crompton, Samuel Willard. *Discovering Ancient Rome*. New York, NY: Britannica Educational Publishing, 2015.

Faulkner, Nicholas, and Erik Gregersen. *The History of Mathematics*. New York, NY: Britannica Educational Publishing, 2018.

Gifford, Clive, Susan Kennedy, and Philip Parker. *Science Year by Year: A Visual History, from Stone Tools to Space Travel*. New York, NY: DK Publishing, 2017.

Hosein, Ann. *Technology of the Ancient World*. New York, NY: Rosen Publishing, 2016.

Idzikowski, Lisa. *The Rise and Fall of the Roman Empire*. New York, NY: Rosen Publishing, 2017.

James, Simon. *Ancient Rome*. New York, NY: DK Publishing, 2015.

Nardo, Don. *Roman Roads and Aqueducts*. San Diego, CA: ReferencePoint Press, 2015.

Pamplona, Alberto Hernández. *A Visual History of Science and Technology*. New York, NY: Rosen Publishing, 2017.

Priwer, Shana, and Cynthia Phillips. *Ancient Monuments*. Armonk, NY: Sharpe Focus, 2009.

BIBLIOGRAPHY

Andrews, Evan. "8 Reasons Why Rome Fell." History, January 14, 2014. https://www.history.com/news/8-reasons-why-rome-fell.

Blitz, Matt. "A New Recreation Shows How Ancient Romans Lifted Wild Animals Into the Colosseum." Smithsonian.com, June 6, 2015. https://www.smithsonianmag.com/travel/how-ancient-romans-got-wild-animals-colosseum-180955580.

Calendars through the Ages. "Early Roman Calendar." WebExhibits. Retrieved April 18, 2019. http://www.webexhibits.org/calendars/calendar-roman.html.

Cartwright, Mark. "Roman Architecture" and other entries. Ancient History Encyclopedia. https://www.ancient.eu.

Chepkemoi, Joyce. "Largest Empires In Human History By Land Area." WorldAtlas, May 12, 2017. https://www.worldatlas.com/articles/largest-empires-in-human-history-by-land-area.html.

Cicero, Marcus Tullius. *Cicero's Three Books of Offices*. Translated by Cyrus R. Edmonds. London, UK: Geroge Bell & Sons, 1874.

Cilliers, L., and F. P. Retief. "The Evolution of the Hospital from Antiquity to the End of the Middle Ages." Curationis, 25, no. 4 (Nov. 2002): 60–66. https://curationis.org.za/index.php/curationis/article/download/806/743.

Claude Moore Health Sciences Library. "Etruscan and Roman Medicine." University of Virginia. Retrieved March 28, 2019. http://exhibits.hsl.virginia.edu/antiqua/etruscan.

BIBLIOGRAPHY

Diederich, Silke. "Columella." Oxford Bibliographies, February 25, 2016. http://www.oxfordbibliographies.com/view/document/obo-9780195389661/obo-9780195389661-0203.xml.

Edmondson, Jonathan. "Metallurgy, Roman." In Oxford Classical Dictionary. Oxford Research Encyclopedias, March 2016. http://oxfordre.com/classics/view/10.1093/acrefore/9780199381135.001.0001/acrefore-9780199381135-e-7260.

Elliott, Jane. "The Romans Carried Out Cataract Ops." BBC News, February 9, 2008. http://news.bbc.co.uk/2/hi/health/7194352.stm.

Haddad, Naif, and Talal Akasheh. "Vitruvius and Ancient Theatres." Cultural Technologies. Retrieved March 29, 2019. http://cultech.net/sites/default/files/Vetruvius.pdf.

Hansen, Roger D. "The Roman Flour Mill at Barbegal." WaterHistory.org. Retrieved March 27, 2019. http://www.waterhistory.org/histories/barbegal.

Hansen, Roger D. "Water Wheels." WaterHistory.org. Retrieved March 27, 2019. http://www.waterhistory.org/histories/waterwheels.

Hays, Jeffrey. "Mining and Resources in Ancient Rome." Facts and Details, October 2018. http://factsanddetails.com/world/cat56/sub408/item2050.html.

Holmes, Tao Tao. "Romans Once Filled the Colosseum with Water and Staged an Epic Mock Sea Battle." Atlas Obscura, January 27, 2016. http://www

.atlasobscura.com/articles/ancient-rome-once-filled-the-colosseum-up-with-water-and-staged-epic-mock-sea-battle.

Klein, Christopher. "1,500 Years Later, Killer Animal Elevator Returns to Colosseum." History, June 9, 2015. https://www.history.com/news/1500-years-later-killer-animal-elevator-returns-to-colosseum.

Labate, Victor. "Roman Engineering" and other entries. Ancient History Encyclopedia. https://www.ancient.eu.

McManus, Barbara F. "Roman Abacus." Retrieved March 22, 2019. http://www.vroma.org/~bmcmanus/abacus.html.

Museum of Ancient Inventions. "Thunder-Making Machine, Egypt, Greece, 100 CE." Smith College. Retrieved March 29, 2019. https://www.smith.edu/hsc/museum/ancient_inventions/hsc19b.htm.

National Geographic Society. "Traces of Ancient Rome in the Modern World." National Geographic, July 6, 2018. http://www.nationalgeographic.org/news/traces-ancient-rome-modern-world.

PBS. "Dome Basics." Retrieved March 22, 2019. https://www.pbs.org/wgbh/buildingbig/dome/basics.html.

Petrovska, Biljana Bauer. "Historical Review of Medicinal Plants' Usage." *Pharmacognosy Reviews* 6, no. 11 (2012): 1–5. PubMed Central, doi:10.4103/0973-7847.95849.

Ralls, Katie M. "Crop Rotations Have Been Around Since Roman Times." Central Oregon Agriculture Research and Extension Center, May 2013. http://oregonstate.edu/dept/coarc/sites/default/files/may_2013_article.pdf.

Santacroce, Luigi, Lucrezia Bottalico, and Ioannis Alexandros Charitos. "Greek Medicine Practice at Ancient Rome: The Physician Molecularist Asclepiades." *Medicines*. 4, no. 4 (2017). doi:10.3390/medicines4040092.

Sleeswyk, André Wegener. "Vitruvius' Odometer." *Scientific American* 245, no. 4 (1981): 188–201.

Stephenson, Steve. "The Roman Hand-Abacus." Ryerson University, September 14, 2014. https://www.ee.ryerson.ca/~elf/abacus/roman-hand-abacus.html.

Thibodeau, Philip. "Greek and Roman Agriculture." In *A Companion to Science, Technology, and Medicine in Ancient Greece and Rome*, edited by Georgia L. Irby, 519–532. Hoboken, NJ: John Wiley & Sons, 2016.

Trentinella, Rosemarie. "Roman Glass." Heilbrunn Timeline of Art History, October 2003. https://www.metmuseum.org/toah/hd/rgls/hd_rgls.htm.

Wasson, Donald L. "The Extent of the Roman Empire." *Ancient History Encyclopedia*, January 5, 2016. https://www.ancient.eu/article/851/the-extent-of-the-roman-empire.

Weisstein, Eric W. "Roman Numerals." MathWorld. http://mathworld.wolfram.com/RomanNumerals.html.

INDEX

A

abacus, 20–21
amphitheaters, 6, 57, 61, 63
  famous, 57–60
Appian Way, 25–26
aqueducts, 6, 13, 32, 33, 34, 35, 37, 45, 56, 60, 63
arcades, 35
archaeologists, 19, 59, 60
arches, 6, 32–34, 35, 38, 39, 55, 58, 59, 61
Archimedes, 29, 38
architects, 6, 28, 37, 55
architecture, 6, 32, 37–38, 61
armies, 9, 10, 13, 14, 22, 25, 49, 51
  importance of, 11
armor, 13, 15
Asclepiades, 49
Asia Minor, 9, 53
astrology, 16, 17–18, 37
astronomy, 16, 17, 19, 37, 62, 63
Augustus, 9–10, 13, 22, 32

B

backfill, 28
ballista, 14
baths, 6, 32–33, 35, 49, 54–56
Baths of Caracalla, 56
bricks, 31, 32–34, 37, 38, 56
bridges, 6, 28, 32, 33, 34, 35, 63

C

Caesar, Julius, 9, 19
calendar, 19
  Egyptian, 19
  farm, 43
  Gregorian, 19
  Julian, 19
  lunar, 18
  solar, 19
capstans, 59–60
castra, 27
Cato the Elder, 51, 62
Colosseum, 57–59, 60
Columella, 43
crops, 35, 40, 42
  diseases, 62
  rotation of, 63

yields, 6, 43

D

ditches, 27, 30, 42
divinities, 53–54
doctors, 6, 49, 53, 54
domes, 6, 32, 38–39, 55
dross, 13

E

emperors, 4, 10
  Augustus, 9–10
  Hadrian, 38
  Marcus Aurelius, 53
  Nero, 51
  Octavian, 9
  Septimius Severus, 56
  Titus, 59
  Trajan, 34
engineers, 5, 6, 11, 12, 15, 21, 27, 28, 32, 37, 56, 63
Etruscans, 7, 22, 50
  customs of, 18

F

farms, 4, 7, 9, 40–42, 44–46
  calendars for, 43
  techniques and technology, 40, 42, 43, 63
fertilizer, 42, 63
flour, 44, 45
forums, 10, 54, 56

G

geometry, 6, 37
gladiators, 53, 59
  games, 57, 59
glass, 50–51
grain, 40, 43, 44
  mills, 6
Greeks, 5, 16–17, 60
  architecture, 37
  astronomer, 19
  conquering of, 9
  gods, 9
  inventions of, 6, 15, 19, 22, 29, 32, 44, 55–56
  mathematician, 38
  physicians, 47, 49, 53
  slaves, 6, 9
  stadiums, 57
groma, 30

H

horoscopes, 18
hospitals, 49
husbandry, animal, 62

hydraulics, 37
hypogeum, 59

I

injuries, battlefield, 49
invaders, 4

K

keystone, 34

L

lamps, oil, 22
legacy, of Rome, 6, 35, 61, 63
legionnaires, 11, 13, 27

M

mathematicians, 6, 29, 38
mathematics, 5, 6, 21, 22, 29, 37, 38, 62
medicine, 6, 37, 47, 49, 50, 51, 53, 54, 63
metals, 12, 13, 15, 29
mills, 6, 35, 44–45, 63
mining, 6, 12, 13, 63
mortar, 13, 32, 34
mortar and pestle, 45
mosaics, 50, 55

N

numerals, Roman, 22, 24

O

oculus, 38, 39
odometer, 28–29
olives, 40, 42, 45–46
onager, 14, 15
organ, 60

P

Pantheon, 38, 39
philosophies, 5, 9, 37, 53
physicians, 47, 49
　famous, 51, 53
physics, 6, 63
piers, 34, 39
pipes, 35, 37, 56, 60
planets, 16, 17
plants, healing, 6, 51, 53
plebeians, 7
Pliny the Elder, 47, 51, 62
plumb lines, 30
population
　of city of Rome, 4
　of republic, 7
Ptolemy, 17
public health, 47, 54

INDEX

importance of baths, 54–56

R

reservoirs, 35, 37, 45, 60
Riotinto Mines, 12
roads, 6, 25, 33, 63
   constructing, 30–31
   features of, 27–28
   leading to Rome, 25–27
Roman Republic, 4, 7, 18, 50
   rise and fall of, 7, 9

S

settling tanks, 35
sewers, 35, 37, 56
sky, 16, 18, 19
slaves, 4, 9, 12, 13, 42, 49, 59
   labor of, 42
   as tutors, 6, 9
Sosigenes, 19
sundials, 19, 22, 37

T

taxes, 10
temples, 10, 32, 38–39, 61, 63

time, 16, 18–19
timekeeping devices, 22
torsion, 15
tufa, 31, 32
tunnels, 13, 28, 31, 35

V

Vitruvius, 28, 29, 37–38, 62

W

wagons, 27, 28
wooden piles, 31

ABOUT THE AUTHOR

Xina M. Uhl discovered her love of history while still in grade school. She went on to obtain a master of arts in history degree from California State University, Northridge, with a focus on the ancient Mediterranean. After teaching college-level history, she moved into educational writing. She has authored books, textbooks, teacher's guides, lessons, and assessment questions in the field of history. When she is not writing, she enjoys travel, photography, and hiking with her dogs. Her blog features her travel adventures and latest fiction projects at http://xuwriter.wordpress.com.

PHOTO CREDITS

Cover © www.iStockphoto/jgaunion; p. 5 Classic Image/Alamy Stock Photo; p. 8. Leemage/Universal Images Group/Getty Images; p. 11 The Metropolitan Museum of Art, New York, Rogers Fund, Transferred from the Library, 1941. www.metmuseum.org/art/collection/search/352200; p. 14 Universal History Archive/Universal Images Group/Getty Images; p. 17 Universal Images Group/Getty Images; pp. 20–21 Science & Society Picture Library/Getty Images; p. 23 prizma/Shutterstock.com; p. 26 De Agostini/S. Vannini/Getty Images; p 29 De Agostini Picture Library/Getty Images; p. 33 NurPhoto/Getty Images; p. 36 Manfred Mehlig/Corbis/Getty Images; pp. 39, 58 © Xina M. Uhl; pp. 41, 46 DEA/G. Dagli Orti/De Agostini Picture Library/Getty Images; p. 43 Peter van Evert/Alamy Stock Photo; p. 48 adoc-photos/Corbis Historical/Getty Images; p. 52 Dioscorides, De medicinali material/Wellcome Collection/CC BY 4.0; p. 55 Anders Blomqvist/Lonely Planet Images/Getty Images; p. 61 Popperfoto/Getty Images; cover and interior pages (dark textured background) Midiwaves/Shutterstock.com; interior pages (scroll pattern page borders) Megin/Shutterstock.com, (yellow marbled page borders) Chizhovao/Shutterstock.

Design and Layout: Nicole Russo-Duca; Editor: Siyavush Saidian; Photo Researcher: Cindy Reiman